Unleashing the Metaverse

I0410292

A
Revolutionary Frontier
in
Chemistry
and
Related Fields

DR. MANOJ BALI

Prof. and Dean, University School of Sciences
Rayat Bahra University, Mohali, Punjab, India
drmanojbali@gmail.com

Preface

In "Unleashing the Metaverse: A Revolutionary Frontier in Chemistry and Related Fields," we embark on an extraordinary exploration of the convergence between virtual reality and the fascinating world of chemistry. This groundbreaking book delves into the revolutionary potential of the metaverse as it intersects with the complexities of chemical research, education, and innovation. Through its immersive virtual landscapes and interactive simulations, the metaverse offers a transformative platform for unleashing the true power of chemistry, pushing the boundaries of discovery, and inspiring a new era of collaboration and understanding. Join us on this captivating journey as we navigate the uncharted territories of the metaverse and witness how its integration with chemistry and related fields unlocks a wealth of opportunities for a brighter future. Welcome to a world where the virtual and the chemical realms intertwine, paving the way for unprecedented advancements and discoveries that will shape the course of science and society alike.

The book delves into the diverse and innovative applications of the metaverse in the field of chemistry and related disciplines. It explores how the metaverse is revolutionizing the way chemistry is studied and taught, offering immersive and interactive experiences for students and educators alike. The book also investigates the metaverse's impact on drug discovery, bio-chemistry, polymers chemistry, food science, environmental science, forensic science, and other specialized fields of chemistry research, enabling scientists to simulate and analyze complex chemical processes in virtual environments. Moreover, it explores how the metaverse is being utilized in fuel research, paint technology, cosmetics development, analytical applications, agriculture, and even in averting accidents, with virtual simulations providing valuable insights and risk assessments. The book highlights the metaverse's transformative potential in advancing the boundaries of chemistry-related fields, paving the way for exciting advancements and discoveries in the future.

Dr. Manoj Bali
Author

Contents

1. Introduction

In the ever-evolving landscape of digital innovation, the concept of the "metaverse" has emerged as a tantalizing vision of a boundless and interconnected virtual reality. The metaverse represents a convergence of immersive technologies, artificial intelligence, and online communities, creating a dynamic and interactive digital universe that blurs the lines between the physical and the virtual. Within this vast and interconnected realm, users can explore, interact, and collaborate in ways that transcend the limitations of our physical world. As we journey into the metaverse, we embark on an exhilarating adventure where imagination knows no bounds, where new horizons of creativity and possibility unfold before our eyes. This exploration will delve into the very essence of the metaverse, its potential applications across various industries, and the profound impact it could have on the way we live, work, and connect. Welcome to the metaverse—a revolutionary frontier that beckons us to dream, to create, and to redefine the very fabric of reality.

The term "Metaverse" refers to a collective virtual shared space where users can interact, socialize, and engage in various activities using immersive technologies, such as virtual reality (VR) and augmented reality (AR). It is a virtual world that exists parallel to the physical world and is accessible through digital devices like VR headsets, smartphones, and computers.

In the subsequent pages we are going to explore the depths of the same and try to visualise how this concept is likely to affect our future, from the point of view of chemistry and otherwise.

2. Key Characteristics of the Metaverse

The metaverse, a concept that blends the physical and digital worlds into an interconnected virtual reality, boasts a myriad of key characteristics that define its immersive and transformative nature. As we embark on this exploration, let us delve into some of the fundamental characteristics that shape the metaverse:

i. **Immersive Experience:** An immersive experience is like stepping into a whole new world, where you feel fully surrounded and engaged by what you see, hear, and feel. It's like being inside a story or a game, where you become an active part of the adventure. Imagine putting on special goggles or using a device that makes you feel like you're inside a movie, video game, or even a faraway place you've never been to before. Everything around you feels so real and interactive that you forget about the real world for a moment and get lost in the excitement of the experience. It's like living in a different reality where you can explore, create, and be a part of something amazing. An immersive experience transcends the boundaries of the ordinary, transporting individuals into captivating digital realms that evoke a profound sense of presence and engagement. It is an invitation to step beyond the confines of reality, to explore, interact, and connect in ways previously unimaginable. Through the magic of cutting-edge technologies such as virtual reality (VR) and augmented reality (AR), an immersive experience envelops the senses, blurring the lines between the physical and the digital. In these captivating digital landscapes, users become active participants, shaping their journey and interacting with the virtual world in real-time. Whether it's diving into a 360-degree interactive story, navigating virtual environments, or collaborating with others in shared spaces, the immersive experience holds the potential to ignite the spark of wonder, creativity, and discovery within us. As we venture further into the realm of immersive experiences, we open the door to a new era of human connection, learning, and expression, where the boundaries of possibility are limited only by our imagination.

The Metaverse provides a highly immersive experience, where users can feel as if they are physically present in the virtual environment. This is achieved through realistic 3D graphics, spatial audio, and interactive elements.

ii. **Shared Environment:** In the context of the metaverse, a shared environment refers to a virtual space where multiple users can interact and collaborate in real-time. It is a digital realm that transcends physical limitations, allowing people from different locations to come together and share a common virtual space as if they were present in the same physical location. In a shared environment, participants can navigate, explore, and communicate with one another, just like in a real-world setting. They can see and interact with each other's avatars or digital representations, engage in conversations, and collaborate on various activities, such as virtual events, meetings, classes, games, or creative projects. The shared environment fosters a sense of social presence and interconnectedness, as users can see and hear each other, making interactions feel more authentic and meaningful. It enables real-time communication, collaboration, and the exchange of ideas, creating a dynamic and interactive space where people can connect, learn, and share experiences.

Shared environments in the metaverse play a pivotal role in facilitating remote collaboration, breaking down geographical barriers, and providing a platform for people to come together in a way that was not possible in the physical world. As the metaverse continues to evolve, shared environments are becoming an essential element in shaping how people interact, socialize, and work in the digital realm. Unlike traditional video games or applications, the Metaverse is a shared space where multiple users can interact with each other in real-time. Users can collaborate, communicate, and socialize within this virtual world.

iii. **Persistent and Evolving:** The metaverse, by its very nature, is both persistent and evolving, offering a dynamic and enduring digital landscape that continually adapts to the needs and desires of its inhabitants. In this multifaceted realm, virtual spaces, objects, and experiences persist over time, ensuring continuity and allowing users to revisit and build upon their past interactions. From the grandest of virtual cities to the tiniest details within them, every creation finds a place within the metaverse's fabric, contributing to its ever-expanding tapestry. The persistent and evolving nature of the metaverse brings forth a sense of continuity and novelty, offering users a space that feels both familiar and full of surprises. With each visit, they find an ever-changing landscape that beckons exploration and offers fresh opportunities for creativity and collaboration. As the metaverse continues to unfold, it is this perfect blend of persistence and evolution that will propel it into a realm of limitless possibilities, forever transforming the way we interact, learn, and connect in the digital age. Moreover, the metaverse is an ever-changing canvas, where creativity knows no bounds. It thrives on the collective imagination of its users, as they actively contribute to its evolution. As new technologies emerge and innovations take shape, the metaverse transforms, incorporating cutting-edge advancements and pushing the boundaries of what is possible. User-generated content, dynamic simulations and real-time interactions ensure that the metaverse remains a living entity, constantly growing and reshaping in response to the desires and endeavors of its inhabitants.

The Metaverse is persistent, meaning it exists and evolves continuously, even when users are not actively present. It allows for ongoing interactions and contributions from users, leading to a dynamic and ever-changing virtual ecosystem.

iv. **Interconnectivity:** The Metaverse is not limited to a single platform or application. Instead, it consists of interconnected virtual spaces and experiences that can be accessed through various devices and applications. The metaverse, at its core, is a sprawling network of interconnectivity, where virtual spaces, experiences, and users seamlessly converge to form a vast and unified digital universe. In this interconnected realm, the barriers of distance and physical limitations dissolve, allowing individuals from all corners of the world to come together, collaborate, and share in a collective digital experience. Through its intricate web of interconnected virtual environments, the metaverse fosters a sense of belonging and social cohesion. Users can navigate from one virtual space to another, exploring diverse landscapes, attending events, and engaging in shared activities with others. This interconnectivity enriches the experience, creating a dynamic and ever-changing virtual ecosystem that thrives on the collective interactions and contributions of its inhabitants.

The metaverse's interconnectivity holds the promise of breaking down traditional barriers and facilitating global collaboration on an unprecedented scale. As this digital frontier continues to expand, the interconnectivity it offers will foster new forms of innovation, communication, and shared experiences, forever transforming the way we connect and interact in the digital age. Embrace the metaverse's interconnectivity, and step into a world where limitless possibilities await at every virtual intersection.

v. **Diverse Applications:** The metaverse boasts a vast array of diverse applications that extend beyond the realms of entertainment and gaming. Its transformative capabilities are evident across numerous industries and fields. In education, the metaverse opens doors to immersive and interactive learning experiences, where students can explore virtual classrooms, conduct simulated experiments, and collaborate with peers worldwide. Businesses and organizations find value in virtual conferencing and collaboration, enabling seamless global interactions and remote work possibilities. In healthcare, the metaverse offers therapeutic applications, such as virtual therapy sessions and medical training simulations. Architects and designers utilize the metaverse to visualize and showcase virtual prototypes of buildings and spaces. Artists showcase their creative expressions in virtual galleries, while social networking takes on new dimensions with shared virtual spaces fostering global connections. Travel and tourism are reimagined as people can virtually explore destinations and landmarks from anywhere. Moreover, environmental research and scientific studies leverage the metaverse to visualize complex data and conduct simulations. With each diverse application, the metaverse unveils its potential to revolutionize industries, redefine experiences, and shape the way we interact in the digital age.

vi. **Virtual Economies:** Virtual economics refers to the study and management of economic activities that take place within virtual environments, such as the metaverse. In these digital realms, users engage in buying, selling, trading, and creating virtual goods, services, and assets that hold value within the context of the virtual world. Virtual economies operate independently from traditional physical economies and are driven by the interactions and decisions of its users. The metaverse plays a pivotal role in virtual economies by providing the platform and infrastructure for these economic activities to thrive. It enables the creation and exchange of virtual goods, such as virtual real estate, digital collectibles, virtual currencies, and in-game items. As users actively contribute to the metaverse by creating content and participating in its ecosystem, they contribute to the growth and dynamism of its virtual economy. The metaverse facilitates secure and transparent transactions, ensures scarcity and rarity of digital assets, and fosters a sense of ownership and value, allowing virtual economies to flourish in ways that mirror real-world economic principles. As the metaverse continues to evolve, its role in virtual economies is expected to expand, offering new opportunities for businesses, entrepreneurs, and content creators to participate in an interconnected and vibrant digital marketplace.

vii. User-Generated Content: User-created content plays a central and indispensable role in contributing to the metaverse's growth, diversity, and overall appeal. It empowers individuals to actively participate in shaping and enriching the virtual world, resulting in a more immersive and engaging experience for all users. Here are some key ways user-created content contributes to the metaverse:

Enriching the Virtual Environment: Users can design and build virtual objects, structures, and environments within the metaverse. These creations add depth and realism to the digital world, transforming it into a dynamic and captivating space. For instance, users can construct virtual houses, theme parks, art installations, or historical landmarks, providing a rich tapestry of places to explore and interact with.

Promoting Creativity and Self-Expression: User-created content fosters creativity and allows individuals to express themselves in novel and imaginative ways. Whether it's designing unique avatars, composing original music, or developing interactive games, users can showcase their talents and interests, contributing to a diverse and vibrant metaverse ecosystem.

Expanding Content Variety: With user-generated content, the metaverse can cater to a wide range of interests and preferences. Users can introduce niche content and experiences that might not have been included in the metaverse's initial offerings. This variety ensures that there is something for everyone, making the metaverse more inclusive and appealing to a broader audience.

Community Building and Engagement: The creation of user-generated content often involves collaboration and shared interests. As users work together on projects, they form communities centered around specific themes, skills, or goals. This fosters a sense of belonging and social interaction within the metaverse, encouraging users to come back and engage more actively.

Continuous Evolution: User-created content ensures that the metaverse is continually evolving. New ideas, innovations, and improvements introduced by users lead to a dynamic and ever-changing virtual world. This constant evolution keeps the metaverse fresh and exciting, enticing both new and returning users to explore and participate.

Economic Opportunities: In some metaverses, user-generated content can have economic implications. Users might be able to monetize their creations through virtual marketplaces or by offering unique experiences and services. This economic aspect motivates creators to invest time and effort in producing high-quality content, further enriching the overall metaverse experience.

Overall, user-created content is a driving force that empowers individuals to take an active role in shaping the metaverse according to their preferences, interests, and talents. By contributing their creations and ideas, users add value to the metaverse ecosystem, making it a more immersive, diverse, and enjoyable virtual realm for everyone.

The concept of the Metaverse has gained significant attention in recent years, driven by advancements in VR and AR technologies, as well as the increasing desire for more immersive and interconnected virtual experiences. Companies and developers are actively exploring the potential of the Metaverse, and it has become a topic of interest in various industries, including entertainment, communication, education, and business. As technology continues to advance, the Metaverse is expected to evolve into an even more integral part of the digital landscape, reshaping how we interact and engage with virtual environments.

Experiences and expertise derived from continuum of technologies are helping metaverse to emerge. We are expecting restructuring of how people think, interact and work with each other in a world blended with digital and physical worlds. As a crude example any metaverse enabled person expected to visit a site will not have to think about how to reach the site, about the environment of the site pollution he has to face at the site, he/she will just be thinking about intellectual inputs he/she is expected to provide.

3. Journey of Metaverse

Novel "Snow Crash" in 1992 by a cyberpunk writer, Neal Stephenson is probably the first instance where the word **"metaverse"** is known to have been used. But since long before that, in Indian folklore and mythology, the concepts of **Maya and Mayavi Nagri** meaning **illusion** have been in use, wherein similar experiences have been painted.

The metaverse is coarsely defined an immersive, three-dimensional (3D), virtual and multi-user online environment. For over 20 years second life has been a virtual world platform.

Over the decades, educational researchers have used this term to describe how learners engage and socialise in the metaverse using digital technologies such as augmented reality (AR), virtual reality (VR) and mixed reality. The metaverse is considered as an immersive, three-dimensional (3D), virtual and multi-user online environment. Desktop-based iteration of the metaverse consisted of agent-based social simulation platforms. The prominence of metaverse reached astonishing heights when world's largest online social network Facebook rebranded itself to Meta

The journey of the metaverse and related technologies has been exciting and ever-evolving landscape. At its heart, the metaverse thrives on the convergence of cutting-edge advancements starting from virtual reality (VR) and augmented reality (AR) to blockchain, 3D reconstruction and artificial intelligence (AI), these technologies are at the forefront of shaping the metaverse. Let's take a closer look at each one:

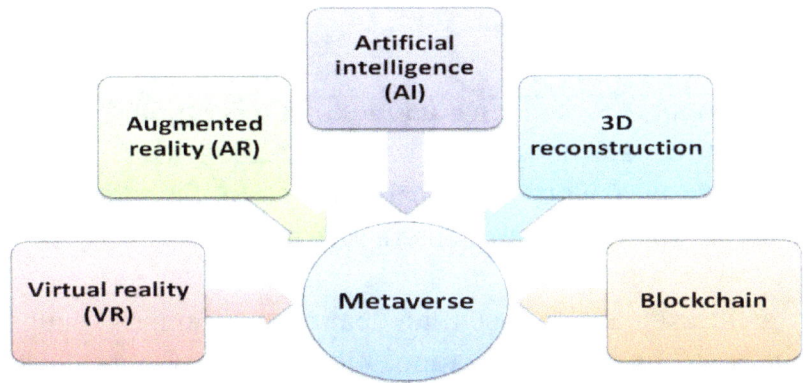

i. **VR:** Virtual reality immerses you in a computer-generated environment, making you feel as if you're physically present in a different world. It's like stepping into a whole new realm, where you can interact with objects and people without actually being there. A popular example of VR is VR gaming. Players wear VR headsets and enter a fully virtual 3D gaming world. They can interact with the environment and objects using handheld controllers, and the game responds in real-time to their movements. For instance, in a VR shooting game, players can physically move, duck, and aim to shoot at virtual enemies. The VR experience provides a heightened sense of immersion and realism, creating an engaging and interactive gaming experience.

Here are some examples of how VR is utilized in various fields:

Gaming: VR gaming allows players to fully immerse themselves in virtual worlds, enhancing the gaming experience. Players can explore, interact with objects, and engage in various activities in a lifelike setting.

Training and Simulation: VR is used for training purposes in fields such as aviation, military, medical, and industrial sectors. Trainees can practice and learn in a safe and

14

controlled virtual environment, reducing real-world risks and costs.

Education: VR is being adopted in classrooms to create interactive and engaging learning experiences. It allows students to explore historical sites, scientific concepts, and complex topics in a more interactive manner.

Healthcare: VR is used for pain management, exposure therapy, and rehabilitation. It helps patients distract themselves from pain during medical procedures and aids in physical and mental therapy sessions.

Architecture and Design: Architects and designers use VR to visualize and present their projects in 3D. Clients can walk through virtual representations of buildings or interior designs before construction starts.

Tourism and Travel: VR provides virtual tours of travel destinations, museums, and cultural landmarks. It allows potential travelers to experience a location before deciding to visit in person.

Social Interaction: Social VR platforms enable people to meet and interact with others in virtual spaces, even if they are physically distant. It creates a more immersive and personal communication experience.

Entertainment and Media: VR is used to enhance entertainment experiences, such as 360-degree videos and virtual concerts, giving users a sense of being present at the event.

Real Estate: VR is used to showcase properties to potential buyers, allowing them to virtually tour houses and apartments from the comfort of their homes.

Psychological Therapy: VR is used in therapeutic settings to treat phobias and anxiety disorders through exposure therapy in a controlled virtual environment.

These examples demonstrate the versatility of VR technology and its potential to revolutionize various industries, making experiences more immersive, interactive, and engaging.

ii. **AR**: Augmented reality overlays digital content onto the real world, enhancing your perception and blending the physical and virtual realms. Think Pokemon Go, where you could see virtual creatures in your real surroundings. Augmented Reality (AR) is a technology that overlays digital content or information onto the real-world environment. Unlike VR, which immerses users in a completely virtual world, AR enhances the real world by adding computer-generated elements to it. AR can be experienced through mobile devices or AR-specific glasses. Examples:

AR Navigation:- An example of AR is AR navigation apps that assist users in finding their way around a city. Imagine you are walking down the street with your smartphone, using an AR navigation app. Through your phone's camera, the app captures the real-world view, and on your screen, you see the live video feed with additional graphical information overlaying it. AR navigation apps can show you arrows or directions on the streets, point out nearby points of interest, and even display reviews or ratings for restaurants and shops that you look at through your phone's camera.

AR Retail Shopping:- AR is also used in retail shopping experiences. With AR retail apps, customers can virtually try on clothing and accessories without physically wearing them. By using their smartphone cameras, customers can see how different outfits will look on them and get a better sense of how the products fit and match their style before making a purchase decision. This enhances the online shopping experience and reduces the need for returns due to incorrect sizing or dissatisfaction with the product's appearance.

In both VR and AR, the applications extend beyond gaming and navigation, finding uses in education, training, design, entertainment, healthcare, and various other industries, providing users with new and innovative ways to interact with the digital and physical worlds.

iii. **Blockchain**: Blockchain technology plays a crucial role in the metaverse. It enables secure and decentralized transactions. It ensures transparency, authenticity, and ownership of digital assets within virtual worlds.

Blockchain technology is a distributed and decentralized ledger system that allows multiple parties to record transactions in a secure and transparent manner. Each block in the chain contains a set of transactions, and once added to the blockchain, it becomes immutable, meaning it cannot be altered or deleted. This ensures the integrity and trustworthiness of the data recorded on the blockchain. An example of blockchain technology is: Cryptocurrencies (e.g., Bitcoin): Bitcoin is perhaps the most famous example of blockchain technology in action. It was created in 2009 as the first decentralized cryptocurrency, and it relies on blockchain to record all its transactions. When someone makes a Bitcoin transaction, it is grouped with other transactions into a block. Miners (participants in the network) compete to solve a complex mathematical puzzle to validate the transactions and add the new block to the blockchain. Once the block is added, the transaction becomes part of a permanent and transparent public record. This decentralized nature and transparency are key features of blockchain that ensure trust and security in the cryptocurrency system.

In this example, blockchain serves as the underlying technology for Bitcoin, providing a transparent and tamper-resistant ledger of all Bitcoin transactions ever made.

Supply Chain Management: Blockchain technology can be used to improve supply chain management by providing a transparent and traceable record of the movement of goods and products through each stage of the supply chain. Each time a product changes hands, the transaction is recorded on the blockchain, and all participants in the supply chain have access to the same information.

This transparency and traceability help reduce fraud, counterfeiting, and inefficiencies in the supply chain. For example, a consumer can scan a product's QR code and view the entire journey of the product, from its origin to the retail store, ensuring its authenticity and ethical sourcing.

Smart Contracts: Smart contracts are self-executing contracts with the terms directly written into lines of code. These contracts are stored and executed on a blockchain network, eliminating the need for intermediaries and ensuring the fulfillment of contract conditions in a trustless manner.

For instance, in real estate, a smart contract could be used to automate the transfer of property ownership once the buyer makes the payment. Once the terms are met (e.g., payment received), the smart contract automatically triggers the transfer of ownership without relying on a third party, reducing costs and increasing efficiency.

These are just a few examples of how blockchain technology is being used in various industries to create transparent, secure, and decentralized systems. As the technology continues to evolve, we can expect more innovative applications and use cases to emerge.

iv. AI: Artificial intelligence powers intelligent virtual avatars, chatbots, and virtual assistants that enhance the metaverse experience. They can understand and respond to human interaction, making the virtual world feel more alive. As the metaverse continues to evolve, we can expect even more exciting advancements. Imagine attending virtual concerts, exploring realistic virtual tourism destinations, or collaborating with others in virtual workplaces.

Artificial Intelligence (AI) refers to the simulation of human intelligence in machines that can perform tasks that typically require human intelligence. It involves the development of computer systems that can learn from data, adapt to new situations, and perform tasks without explicit programming. AI encompasses various techniques such as machine learning, natural language processing, computer vision, and robotics. Let's illustrate AI with a simple example of a spam email filter:

Example: Spam Email Filter: Imagine you have an email service that receives a large number of emails every day, and among them, some are spam messages. Your goal is to automatically filter out these spam emails and move them to a separate spam folder, preventing them from reaching the user's inbox.

Traditional Rule-Based Approach: In the past, spam filters were rule-based, where a set of rules was defined to flag emails containing specific keywords or patterns typically found in spam messages. For example, if an email contained words like "free," "prize," "lottery," and excessive use of exclamation marks, it might be flagged as spam.

However, this rule-based approach had limitations. Spammers quickly adapted and found ways to bypass these rules, and legitimate emails containing similar words were also mistakenly flagged as spam.

AI-Based Approach: To overcome the limitations of the rule-based approach, an AI-based spam filter was developed using machine learning techniques.

Data Collection: A large dataset of emails, labeled as spam or non-spam (ham), is collected for training the AI model.
Feature Extraction: The AI model analyzes each email and extracts relevant features, such as the frequency of certain words, presence of links, sender information, and email structure.

Model Training: The AI model uses this labeled dataset to learn patterns and relationships between features and whether an email is spam or not.

Prediction: Once the model is trained, it can predict whether a new, unseen email is spam or ham based on the learned patterns from the training data.

Continuous Learning: The spam filter system can continuously learn from new data and user feedback. If a user marks an email as spam that wasn't flagged by the filter, the system can incorporate this feedback to improve its accuracy over time.

The AI-based spam filter is more effective and adaptable compared to the rule-based approach. It can identify previously unseen spam patterns and reduces false positives (legitimate emails incorrectly flagged as spam) and false negatives (spam emails not caught by the filter).

This example demonstrates how AI, specifically machine learning, can be applied to solve real-world problems, learn from data, and continuously improve its performance, making it a powerful tool in various domains, ranging from email filtering to more complex tasks like autonomous vehicles and medical diagnostics.

v. **3D reconstruction:** 3D reconstruction plays a vital role in enhancing the metaverse experience. By using advanced computer vision techniques, 3D reconstruction technology can create realistic virtual representations of real-world objects, environments, and even people. Imagine being able to explore a virtual world that looks and feels just like the real one. With 3D reconstruction, this becomes possible. It allows for the creation of highly detailed and immersive virtual environments, increasing the sense of presence and engagement within the metaverse. Furthermore, 3D reconstruction enables realistic avatars by capturing and reconstructing human figures. This means that in the metaverse, you can have a digital representation of yourself that closely resembles your physical appearance, enabling more personal and authentic interactions with others. Additionally, 3D reconstruction enhances the metaverse's ability to interact with the physical world. For example, it can be used in architectural design, allowing architects to create virtual models of buildings and test their functionalities before they are built. It can also assist in virtual tourism, enabling people to explore real-world landmarks and historical sites from the comfort of their homes. Overall, 3D reconstruction brings a new level of realism and immersion to the metaverse, making the virtual experience even more captivating and enjoyable. Virtual Reality (VR) and Augmented Reality (AR) both rely heavily on 3D reconstruction to create immersive experiences. In VR, 3D models of environments are created to provide a lifelike experience for users. In AR, 3D models are overlaid onto the real world, enhancing the user's perception of reality.

This technology is widely used in various fields, including computer vision, medical imaging, archaeology, entertainment, and more. Here are some examples of 3D reconstruction applications:

Photogrammetry: This technique involves capturing multiple photographs of an object or a scene from different angles and using specialized software to reconstruct a 3D model. It's commonly used in fields like architecture, archaeology, and cultural heritage preservation.

Medical Imaging: In medical imaging, 3D reconstruction is used to create detailed 3D models of internal organs and structures from various imaging techniques like CT scans, MRI, and ultrasound. These models assist in surgical planning, diagnostics, and medical research.

3D Scanning and Printing: 3D reconstruction is used in 3D scanning to capture the shape and texture of physical objects, which can then be converted into digital 3D models. These models can be edited, modified, or 3D printed for various purposes like rapid prototyping, art, or manufacturing.

Computer Graphics and Animation: In the entertainment industry, 3D reconstruction plays a crucial role in creating lifelike characters, scenes, and special effects for movies, video games, and animations.

Robotics and Autonomous Vehicles: 3D reconstruction is used in robotics and autonomous vehicles to perceive and understand the surrounding environment. It helps robots and vehicles navigate, avoid obstacles, and make informed decisions based on 3D representations of the world.

Architectural Visualization: 3D reconstruction is employed to create realistic 3D models of buildings and environments for architectural visualization. It allows architects and clients to visualize and explore designs before construction begins.

Reverse Engineering: 3D reconstruction is utilized in reverse engineering to create digital representations of existing objects, allowing for analysis, modification, and replication of the original item.

Cultural Heritage Preservation: In the field of cultural heritage, 3D reconstruction is used to digitize and preserve historical artifacts, monuments, and archaeological sites, making them accessible to researchers and the public.

These are just a few examples of how 3D reconstruction is applied across various domains to enhance understanding, creativity, and problem-solving capabilities. The technology continues to advance, enabling new applications and opportunities in the future.

Thus the journey of Metaverse and its related technologies has been a captivating and dynamic one. The Metaverse thrives on the convergence of cutting edge advancements spanning from Virtual Reality (VR) and Augmented Reality (AR) to Blockchain, 3D Reconstruction and Artificial Intelligence (AI). Each of these technologies plays a pivotal role in shaping the Metaverse evolution and potential.

4. Applications of Metaverse

a. Metaverse in Chemistry

Chemistry is the scientific discipline that studies the composition, structure, properties, and reactions of matter. It is often referred to as the central science because it connects and overlaps with other scientific fields like biology, physics, and environmental science. Chemistry delves into the building blocks of the universe, investigating the behavior of atoms, molecules, and ions. Through various experimental and theoretical methods, chemists explore the interactions between different substances, how they transform, and the energy changes involved in these processes. From understanding the intricacies of chemical reactions to designing new materials and drugs, chemistry plays a fundamental role in shaping our world and improving the quality of life. It is a dynamic and ever-evolving field, driving innovations across industries and offering solutions to the challenges facing humanity in areas like energy, medicine, agriculture, and environmental conservation.

The world of chemistry is also about to undergo a groundbreaking transformation with the emergence of the Metaverse. The application of Metaverse technology in chemistry promises to revolutionize research, education, collaboration, and even the way we interact with molecules and chemical processes. In this article, we will explore the vast potential of this immersive virtual world and its impact on the field of chemistry. The major activities which are expected to surface as a result of Metaverse are:

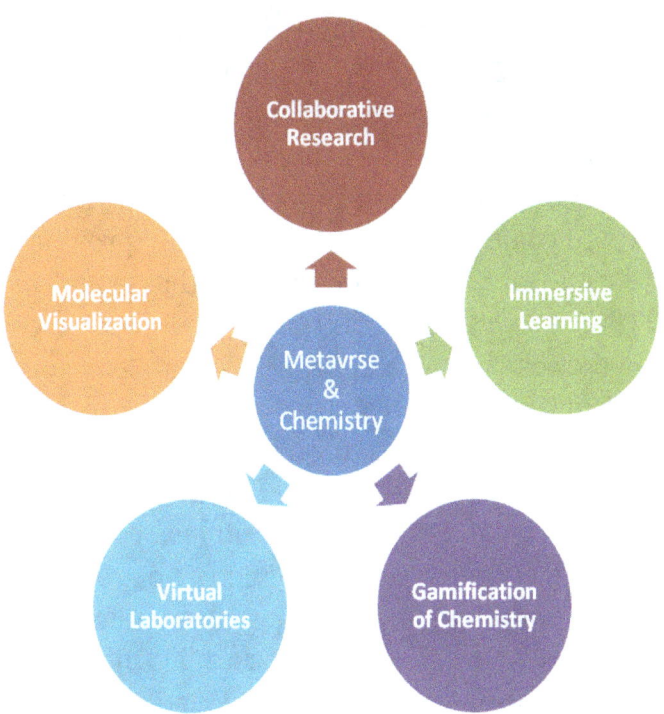

i. Virtual Laboratories: One of the most exciting applications of the Metaverse in chemistry is the creation of virtual laboratories. Researchers can now conduct experiments without the need for physical equipment or hazardous chemicals. By simulating chemical reactions in a virtual environment, scientists can explore various scenarios, optimize reaction conditions, and even predict the outcome of reactions before stepping into a physical lab. This not only saves time and resources but also enhances safety and promotes more efficient experimentation.

The following top virtual lab sites and apps are all free, highly engaging, and informative—and most don't require registration. Since most browsers no longer support Java or Flash, sites built exclusively with those outdated technologies have been excluded.

- **National Science Digital Library: Chem Collective Virtual Labs** : The National Science Digital Library (NSDL) hosts an exceptional platform known as the Chem Collective Virtual Labs, offering an innovative and interactive approach to learning chemistry. As part of the NSDL initiative, Chem Collective provides a series of virtual laboratory simulations that enable students and educators to perform various chemistry experiments and exercises in a safe and virtual environment. These simulations cover a wide range of topics, from general chemistry concepts to more specialized areas, allowing users to explore chemical reactions, molecular interactions, and lab techniques. The Chem Collective Virtual Labs are designed to enhance students' understanding of fundamental chemistry principles and encourage active participation in the learning process. With its user-friendly interface and access to detailed data and analysis, students can practice lab techniques, observe chemical reactions, and apply theoretical knowledge in a hands-on manner. This valuable resource complements traditional laboratory experiences and offers a flexible and accessible learning tool for educators and learners alike, enriching the study of chemistry and promoting STEM education.

- **PBS: Nova Labs**: PBS: Nova Labs is an innovative and educational online platform that brings science, technology, and exploration to life. Developed by the renowned PBS Nova team, Nova Labs offers a diverse collection of interactive tools, games, and activities that engage learners of all ages in scientific exploration and discovery. Through its captivating modules, users can delve into a wide range of topics, from astronomy and biology to physics and engineering. Each lab presents real-world challenges and simulations, encouraging

27

critical thinking and problem-solving skills. Whether it's exploring the mysteries of space, unraveling the complexities of genetics, or understanding the forces of nature, Nova Labs provides an immersive and interactive learning experience that transcends traditional educational methods. The platform's user-friendly design allows educators, students, and lifelong learners to explore scientific concepts in a hands-on manner, fostering a deeper understanding and appreciation of the natural world and the wonders of science. PBS: Nova Labs stands as a testament to the power of digital technology in making science accessible, engaging, and enjoyable for audiences worldwide.

- **University of Colorado Boulder: Interactive Simulations for Science and Math** : it is perhaps the richest single repositories of online science interactives and virtual labs. It houses more than 150 physics, chemistry, math, biology, and Earth science topics. With their engaging and immersive nature, these simulations enable students to explore scientific phenomena and mathematical concepts, providing a dynamic and hands-on approach to understanding complex topics. Available in dozens of languages, the interactives are searchable by grade level, subject, and accessibility.

- **Reactor Lab**: Reactor Lab is educational software developed by the University of California, Berkeley, to help students learn about nuclear reactors and nuclear physics concepts. This interactive tool allows users to simulate and analyze the behavior of nuclear reactors under different conditions. Through Reactor Lab, students can gain practical insights into nuclear science, reactor design, and safety protocols, making it a valuable resource for nuclear engineering and physics education.

o **Line Rider**: Awesome physics simulation/game/lab experiment. Line Rider another popular and addictive online game that allows players to draw and create their own custom tracks for a sledder to ride on. Developed by Boštjan Čadež, Line Rider's simple yet engaging concept has captivated players worldwide since its initial release in 2006. With its user-friendly interface and creative possibilities, Line Rider offers a unique and enjoyable experience for players to design and share their gravity-defying sledding adventures.

o **Explore Learning Free Gizmos**: Gizmos are fascinating and versatile interactive tools that offer immense appeal for exploring mathematical and scientific concepts. By signing up for a free teacher account, you can access a diverse collection of free Gizmos that are regularly updated. Among the current selection of 36 free Gizmos, you'll find explorations on Hubble's Law, the circulatory system, and periodic table trends. Each Gizmo includes multiple manipulatives that cater to various aspects of the subject, along with assessment questions, student guides, and teacher guides. Rest assured that all Gizmos are certified compliant with COPPA, FERPA, and CSPC regulations, ensuring a safe and secure learning environment.

o **Zooniverse**: Can students and teachers with no specialized training participate in real-world research? Yes — and the Zooniverse platform is designed to facilitate just that. Zooniverse is a groundbreaking citizen science platform that engages millions of volunteers in scientific research and exploration. Created by the Citizen Science Alliance, Zooniverse offers a wide array of projects in various fields, such as astronomy, biology, climate science, and more. Participants can contribute their time and expertise to help analyze large datasets, classify celestial objects, identify species, and make significant contributions to scientific discoveries. Zooniverse's collaborative approach to research has transformed how scientists tackle complex challenges, making it an inspiring example of the power of crowd sourcing in advancing scientific knowledge. Current projects run the gamut from the arts to biology to social science; in all, 79 active projects span 11 disciplines. The education guide provides examples and resources for teachers interested in using this powerful research tool in their classrooms. Classrooms can even create their own projects by uploading data to the Zooniverse.

- **The Concord Consortium Free Interactive STEM Activities:** Fine, extensive collection of interactive STEM resources, searchable by grade level, activity type, and STEM topic. Their educational resources are accessible to students, teachers, and learners of all ages, promoting hands-on exploration and inquiry-based learning. With a focus on innovative technology and real-world applications, the interactive STEM activities provide engaging and dynamic learning experiences, empowering individuals to delve deeper into STEM subjects and develop essential skills for the future.

- **PraxiLaBS**: With a free basic account, users can access six simulations in English and Arabic. PraxiLaBS is an immersive and innovative platform that offers virtual laboratories for scientific research and education. Developed by the Centre for Development of Advanced Computing (C-DAC) in India, PraxiLaBS provides a diverse range of virtual experiments across various disciplines, including physics, chemistry, biology, and engineering. The simulated labs are remarkable, offering sophisticated graphics, PDF and animated guides, assessments, and enrichments activities.

o **LabXChange Harvard**: More than 300 free lab simulations that can be easily shared or embedded. LabXChange is an online learning platform developed by Harvard University that offers a wide range of educational resources for science and biology. It provides interactive lab simulations, virtual experiments, and customizable learning pathways, catering to students, educators, and lifelong learners. With its user-friendly interface and access to high-quality scientific content, LabXChange Harvard enhances the study of biology and empowers users to explore and engage with complex scientific concepts in a flexible and dynamic manner. It is rich resource for virtual labs, including some very topical entries, such as Covid-19 infection spread, climate change predictor, and model of hydraulic fracking.

ii. **Molecular Visualization:** Visualizing complex molecular structures and interactions has always been a challenge in chemistry. However, with the Metaverse, researchers can dive into a three-dimensional representation of molecules, enabling a better understanding of their properties and behavior. By interacting with virtual molecules, scientists can explore their structures, study their dynamics, and gain insights into their reactivity. This enhanced molecular visualization provides a powerful tool for teaching, research, and drug discovery efforts.

Here is a list of some other popular molecular visualization software:

- o **MolAR** has been developed that turns hand drawn structures into 3D lets people turn their hand-drawn structures into 3D augmented reality molecules has caused excitement among chemists. **MolAR** is not only instructional but also fun to use. 'The app provides quick visualization of the 3D structure of a molecule in augmented reality (AR), making it effortless for users to understand molecular structures,' (Team leader Todd Martinez of Stanford University, US). 'This offers a far richer learning experience. Rather than static 2D structures on a page, the molecules appear in 3D on their desk.

- o **PyMOL**: PyMOL is a widely used molecular visualization program known for its ease of use and powerful capabilities. It is available for both commercial and open-source versions.

- o **ChimeraX**: ChimeraX is a next-generation molecular visualization program developed by the UCSF Chimera team. It is designed to be highly interactive and user-friendly.

- o **Jmol**: Jmol is an open-source Java-based molecular visualization program that can be embedded into webpages. It is widely used for educational purposes and web-based molecular visualization.

- o **RasMol**: RasMol is a widely used molecular visualization software for the interactive display of molecular structures and animations.

- o **Avogadro**: Avogadro is an open source molecular modeling software that includes visualization and analysis tools for molecular structures.

o **Maestro**: Maestro is acommercial molecular visualization and analysis software developed by Schrödinger. It is part of the Schrödinger Suite, which includes various computational tools for drug discovery and materials science.

These are just a few examples of the many molecular visualization softwares available to researchers and scientists. The choice of software often depends on the specific research needs, budget constraints, and user preferences.

iii. **Collaborative Research:** The Metaverse promotes seamless collaboration among chemists worldwide. Scientists can gather in virtual meeting rooms, exchange ideas, and work together on projects in real-time, regardless of their physical location. This level of collaboration transcends geographical boundaries and time zones, fostering a global community of chemists working towards common goals. The Metaverse serves as a platform for brainstorming, sharing research findings, and collectively tackling complex scientific challenges.

Several companies and projects were exploring the idea of virtual meeting spaces within metaverse-like environments. Here are a few examples:

o **Spatial**: Spatial is a virtual meeting platform that allows users to collaborate in a shared 3D space using VR headsets or web browsers. It enables participants to present, share documents, and interact with 3D objects in real-time.

o **AltspaceVR**: AltspaceVR is a social VR platform where users can meet and interact in virtual spaces. It offers virtual meeting rooms for presentations, discussions, and events.

o **Virbela**: Virbela is a virtual world platform designed for remote work and education. It provides virtual meeting rooms, classrooms, and auditoriums where users can collaborate and interact using avatars.

o **Mozilla Hubs:** Mozilla Hubs is an open-source platform that allows users to create and share virtual spaces for meetings and events. It can be accessed through web browsers or VR headsets.

o **Rumii**: Rumii is a VR collaboration platform that offers virtual meeting rooms for remote teams to work together in 3D spaces. It supports features like file sharing, screen sharing, and voice communication.

o **Engage:** Engage is an education and training platform that provides virtual classrooms and meeting spaces in VR. It is used for workshops, conferences, and virtual events.

o **MeetinVR:** MeetinVR is a VR collaboration platform that offers virtual meeting rooms with features like whiteboarding, file sharing, and spatial audio for a more immersive meeting experience.

Please note that the metaverse concept is continually evolving, and new platforms and technologies may have emerged since compilation of this book.

iv. **Immersive Learning:** Chemistry education is being transformed by the Metaverse. Traditional textbooks and lectures can now be replaced with immersive learning experiences. Students can explore virtual chemistry labs, conduct experiments, and visualize complex concepts in real-time. This interactive learning environment enhances engagement, comprehension, and retention of knowledge. The Metaverse in chemistry education opens up new avenues for students to develop practical skills and gain hands-on experience, ultimately preparing them for future careers in the field.

Gamification of Chemistry: The Metaverse allows for the gamification of chemistry, making the learning process more enjoyable and accessible. Virtual reality (VR) and augmented reality (AR) technologies can be employed to create interactive games that teach chemical concepts, enhance problem-solving skills, and encourage exploration. By combining education with entertainment, the Metaverse makes chemistry more approachable, engaging, and appealing to learners of all ages. Yale Center for Green Chemistry and Green Engineering developed an educational online game, which introduces concepts of sustainability, chemistry and rational design to undergraduate students using a virtual environment.

Here are some potential game ideas that could leverage metaverse technologies for chemistry education:

- **The Safer Chemical Design Game** - Gamification of green chemistry and safer chemical design concepts for students.

  The game's primary objective is to familiarize students with safer chemical design concepts by allowing them to manipulate physicochemical parameters to reduce undesirable biological and environmental interactions of a hypothetical commercial chemical. It simulates the decision-making process used by professionals in designing new chemicals, taking into account real-world constraints that influence product development. Using multi-criteria decision analysis, players must design a safer and more sustainable chemical product by selecting different combinations of molecular parameters affecting toxicity, biodegradability, biotransformation, and overall performance. Throughout the game, players encounter scenarios with increasing complexity, divided into challenges related to the chemical product's attributes.

To guide students, "Tips" provide additional information to aid their parameter choices. The game incorporates both lower order learning objectives (fact memorization) and higher order learning objectives (knowledge transfer) as players optimize the chemical product against multiple requirements. Evaluation is based on function and avoidance of toxicity, with feedback after each task allowing players to redesign the chemical product and improve their design.

The game is designed based on the 12 Principles of Green Chemistry and introduces toxicology and pharmacokinetic concepts. While rooted in scientific principles, it aims to engage and entertain students using game mechanics, including goal orientation, reward points, reinforcement based on feedback, and fun orientation (gamification principles). The game's overall goal, developed by the Yale Center for Green Chemistry and Green Engineering, is to educate students about real-life challenges and encourage them to design safer chemicals effectively.

The game aims to achieve the following objectives:

a. Develop a scientifically accurate game that allows students to solve problems as if they were practicing professionals.
b. Captivate and entertain students through an engaging storyline that relates to real-world issues, particularly sustainability.
c. Instill sustainable design principles and encourage life cycle thinking among students.
d. Educate students about toxicity, biodegradability, and overall performance of a hypothetical chemical product.
e. Integrate systems thinking and interdisciplinary content, exploring the intersection of chemistry, toxicology, and environmental science.

The Safer Chemical Design Game can be accessed at: **http://greenchemistry.yale.edu/education/und ergraduate-graduate** → Safer Chemical Design Game.

o **Qualium Systems:** In this project a VR edutainment app has been developed to help students in studying chemistry with gamification approach. This application contains a learning plan in playful form with animation, interactive environment, and a score table reflecting current progress.

o **Futuclass Education:** Provides environment to learn basic chemistry through engaging gamified experiences in VR, covering topics such as: Atom Structure, Oxygen, Hydrogen, Salts, Acids & Bases, Reaction Balancing, Moles and Molarity. Co-created with educational experts, the learning value of each VR lesson has been proven with real students in classroom and at home. Subscription service for schools includes an online platform to run VR lessons with ease.

o **Virtual Chemistry Lab Simulations:** Create a realistic and immersive virtual chemistry lab in the metaverse, where players can perform various chemistry experiments, reactions, and observations in a safe and interactive environment.

o **Chemistry Quests in Virtual Environments:** Develop chemistry-themed quests in the metaverse, where players must explore different virtual environments to collect elements, molecules, or solve chemistry-related challenges.

o **Interactive Molecular Building Games:** Design a game in the metaverse that allows players to construct complex molecules and chemical structures using interactive 3D modeling tools and simulations.

o **Chemistry Battles or Challenges:** Create competitive chemistry challenges or battles in the metaverse, where players use their chemical knowledge to outwit opponents and win chemistry-themed contests.

o **Virtual Chemistry Olympiads:** Organize virtual chemistry Olympiads in the metaverse, where students from different locations can compete in chemistry quizzes, challenges, and collaborative problem-solving activities.

o **Chemistry Trivia and Quiz Games:** Develop interactive chemistry trivia and quiz games in the metaverse, where players can test their knowledge of chemical elements, reactions, and principles.

Please note that while these examples showcase the potential integration of metaverse technologies with chemistry-based games, the actual existence of such games may vary.

The Metaverse has the potential to transform the field of chemistry in ways we could not have imagined before. From virtual laboratories and immersive molecular visualization to collaborative research and gamified education, the Metaverse offers endless possibilities for advancing chemistry. As this technology continues to evolve, we can expect further innovations and discoveries that will shape the future of chemistry, paving the way for a new era of scientific exploration and understanding.

Examples

Here are a few examples of three-dimensional views of chemical interactions in the Metaverse:

○ Protein-Ligand Interactions: In the Metaverse, researchers can visualize the three-dimensional structure of a protein and its binding with a ligand molecule. This visualization allows them to understand how the ligand interacts with specific amino acids or binding sites on the protein. By manipulating the virtual molecules, scientists can study the strength and nature of these interactions, facilitating the design of more effective and targeted drugs. In the field of biochemistry and drug discovery, understanding protein-ligand interactions is crucial for developing new drugs and therapeutic agents. The metaverse can offer a unique and immersive approach to studying and visualizing these interactions, providing researchers with valuable insights and accelerating the drug discovery process. Here's an example of how the metaverse can be used to explore protein-ligand interactions:

Virtual Protein-Ligand Interaction Exploration in the Metaverse

Objective: To study the binding interactions between a protein target and a potential ligand molecule for drug discovery purposes.

> ➢ Virtual Molecular Visualization: Researchers wearing VR headsets are immersed in a 3D virtual laboratory where they can see and manipulate protein structures and ligand molecules in real-time.
> ➢ Protein Target Selection: The researchers can choose a specific protein target of interest from a library of known drug targets or input their custom protein structure for analysis.
> ➢ Ligand Database Access: A virtual database contains a wide range of ligand molecules, including known drugs and potential candidates. Researchers can select ligands to study their interactions with the chosen protein target.
> ➢ Docking Simulations: Using advanced molecular docking algorithms, researchers can dock the selected ligands onto the protein target to predict their potential binding orientations and interactions.
> ➢ Interactive Binding Analysis: The VR environment enables researchers to visually inspect and analyze the interactions between the protein and ligand. They can observe hydrogen bonds, hydrophobic interactions, and other intermolecular forces that stabilize the complex.

> Energy Calculations: Researchers can perform real-time energy calculations to evaluate the binding affinity and stability of the protein-ligand complex.

> Virtual "Wet Lab" Simulations: In a simulated "wet lab" environment, researchers can virtually perform biochemical assays and test the binding predictions made by the docking simulations.

> Collaborative Research: Multiple researchers can collaborate within the metaverse, working together on the same protein-ligand complex or sharing their findings with others.

> Drug Design Iteration: Based on the results of the virtual interactions and simulations, researchers can modify the ligand's chemical structure and study its impact on binding. This iterative process aids in designing more potent and selective drugs.

Benefits:

a. Enhanced Visualization: The 3D representation of proteins and ligands in the metaverse allows researchers to observe intricate details of the binding interactions, leading to better insights.

b. Efficient Experimentation: Virtual simulations and "wet lab" experiments in the metaverse eliminate the need for physical materials and resources, making the drug discovery process more efficient and cost-effective.

c. Collaboration and Global Access: Researchers from different locations can come together in the metaverse to collaborate and share their expertise, enabling international cooperation on drug discovery projects.

d. Rapid Prototyping: Researchers can quickly test multiple ligands against a given protein target in the virtual environment, accelerating the early stages of drug design.

By leveraging the metaverse for protein-ligand interaction studies, researchers can gain a deeper understanding of molecular interactions, leading to the development of more effective and targeted drugs for various diseases and medical conditions.

Enzyme-Substrate Interactions: The Metaverse offers a unique perspective on enzyme-substrate interactions. Researchers can observe how substrates fit into the active site of an enzyme, allowing them to understand the mechanism of enzyme catalysis. By virtually manipulating the substrate molecules, scientists can study how different chemical groups and functional moieties interact with the enzyme, aiding in the design of enzyme inhibitors or optimizing enzymatic reactions.

Crystal Structures: The Metaverse provides a platform for visualizing crystal structures of various compounds. Researchers can explore the three-dimensional arrangements of atoms within a crystal lattice, allowing them to understand the geometric arrangement and intermolecular forces between molecules. This insight is particularly useful in drug development and materials science, where crystal structures play a crucial role in determining properties and behaviors.

Molecular Dynamics Simulations: Using the Metaverse, scientists can simulate the movement and interactions of molecules in real-time. This dynamic representation provides a comprehensive view of how molecules behave and interact with their surroundings. Researchers can observe conformational changes, molecular vibrations, and even chemical reactions unfolding in three-dimensional space. These simulations help in understanding molecular behavior, optimizing reaction conditions, and predicting the outcomes of chemical processes.

Nanoparticle Interactions: Nanoparticles have unique properties due to their size and surface characteristics. In the Metaverse, researchers can visualize the interaction between nanoparticles and other molecules. This allows for a detailed exploration of how nanoparticles bind to target molecules, how their surface chemistry influences interactions, and how they can be engineered for specific applications such as drug delivery or catalysis. These examples highlight the power of the Metaverse in providing immersive and interactive three-dimensional views of chemical interactions. By leveraging this technology, researchers can gain valuable insights into the behavior of molecules, enabling them to make more informed decisions and advancements in various fields of chemistry.

b. Metaverse in Teaching of Chemistry

Effective teaching is a dynamic process that goes beyond simply conveying information to students. It involves engaging learners, fostering critical thinking, and tailoring instructional methods to meet individual needs. In the modern educational landscape, effective teaching has become more essential than ever. With advances in technology and access to vast amounts of information, students now require not only knowledge acquisition but also the development of crucial skills like problem-solving, creativity, and adaptability. Effective teaching embraces interactive and student-centered approaches, leveraging technology and digital tools to enhance learning experiences. It promotes a growth mindset, encourages collaboration, and incorporates real-world applications to make education relevant and meaningful. Moreover, modern teaching recognizes the diverse backgrounds and learning styles of students, fostering inclusivity and equity in the classroom. By adapting to the changing needs of students and leveraging innovative pedagogies, effective teaching prepares learners to thrive in a rapidly evolving world and become lifelong learners equipped for success in the 21st century.

The metaverse has the potential to revolutionize the way chemistry is taught and learned. Here are some exciting applications of the metaverse in teaching chemistry:

i. **Virtual Laboratories:** The metaverse can provide virtual laboratory experiences where students can safely conduct chemistry experiments in a digital environment. They can manipulate virtual chemicals, observe reactions, and practice lab techniques without the need for physical resources or safety concerns.

ii. **Interactive Molecular Visualization**: Chemistry concepts often involve complex molecular structures and interactions. The metaverse can offer immersive 3D visualizations of molecules, allowing students to explore and interact with them in real-time. This enhances understanding and enables students to visualize concepts that are difficult to grasp in traditional learning settings.

iii. **Collaborative Learning**: The metaverse can facilitate collaborative learning experiences for chemistry students. They can join virtual classrooms, work on group projects, and interact with peers and instructors in real-time. This promotes active engagement, discussion, and knowledge sharing among students.

iv. **Gamified Learning Experiences:** By integrating game elements into chemistry education within the metaverse, students can have fun while learning. Gamified experiences can include solving chemistry puzzles, participating in virtual chemistry quests, or even engaging in virtual chemistry competitions. This approach enhances motivation and makes learning chemistry more enjoyable.

v. **Virtual Field Trips**: The metaverse can transport students to various chemical environments, such as industrial plants, research laboratories, or natural settings where chemical reactions occur. Virtual field trips provide immersive experiences, allowing students to explore chemical phenomena in different contexts and deepen their understanding.

vi. **Remote Learning and Accessibility**: The metaverse enables remote learning opportunities, making chemistry education accessible to students who may not have access to physical laboratories or specialized equipment. It promotes inclusivity and allows students to engage in chemistry education from anywhere in the world. These applications of the metaverse in teaching chemistry have the potential to enhance student engagement, understanding, and overall learning outcomes. It's an exciting space where technology and education intersect to create immersive and interactive learning experiences.

c. **Metaverse in Drug Discovery**

Drug discovery is a complex and vital process in the field of pharmaceutical research and development. It involves the identification and development of new medications to treat diseases and improve human health. The process typically begins with target identification, where researchers identify specific molecules or biological pathways involved in a disease. Subsequently, high-throughput screening and virtual screening techniques are employed to identify potential drug candidates from vast compound libraries. Once potential compounds are identified, they undergo various preclinical tests to assess their safety and efficacy. Promising candidates then advance to clinical trials, where their effectiveness and safety are evaluated in human subjects. The drug discovery process is time-consuming, expensive, and carries a high risk of failure, but successful discoveries have the potential to revolutionize healthcare and save lives by providing new treatments and therapies for a wide range of medical conditions.

The future of the metaverse in drug discovery is incredibly exciting! Imagine a virtual world where scientists and researchers can collaborate and explore vast digital landscapes to discover new drugs and treatments. In the metaverse, they can simulate and experiment with molecules and their interactions, accelerating the drug discovery process. Imagine virtual laboratories, where AI-powered avatars assist scientists in designing and testing new compounds, analyzing data in real-time, and predicting their effectiveness. Researchers from around the world can gather in this digital realm, breaking down geographical barriers and fostering collaboration in ways never seen before. Moreover, the metaverse can also make drug discovery more accessible to everyone. It could allow patients to actively participate in clinical trials by virtually experiencing potential treatments and providing valuable feedback in a safe and controlled environment. The potential applications of the metaverse in drug discovery are endless. It could revolutionize the way we develop and test new drugs, leading to faster and more effective treatments for various diseases. It would be fantastic to dive into a virtual world of scientific exploration and innovation!

Here, we will explore the application of the Metaverse in the field of drug discovery. Traditional drug development processes involve a significant amount of time, resources, and experimentation. However, the Metaverse offers an innovative approach to accelerate the drug discovery process by leveraging immersive virtual environments and collaborative tools. Scenario: A pharmaceutical company, ABC Pharma, is aiming to develop a novel drug to target a specific disease. Traditionally, this process involved multiple steps, including laboratory experiments, compound screening, and clinical trials. However, ABC Pharma decides to leverage the power of the Metaverse to streamline their drug discovery efforts.

i. **Virtual Laboratory Simulation:** Using the Metaverse, ABC Pharma creates a virtual laboratory where their researchers can simulate and conduct experiments virtually. They can design and test different chemical compounds, predict their properties and interactions, and analyze potential drug-target interactions. This virtual laboratory eliminates the need for physical prototypes and allows for rapid iteration and optimization of drug candidates.

ii. **Collaborative Research**: ABC Pharma connects chemists, biologists, and other experts from around the world in a virtual collaboration space within the Metaverse. This enables real-time collaboration and knowledge sharing, transcending geographical limitations. Researchers can discuss findings, share data, and collectively analyze and refine drug candidates. The immersive environment of the Metaverse promotes a sense of presence and collaboration, accelerating the decision-making process. Molecular Visualization: To enhance the understanding of molecular structures, ABC Pharma utilizes the Metaverse's advanced visualization tools. Researchers can explore three-dimensional representations of drug molecules, analyzing their structures, identifying potential binding sites, and predicting their behavior within the human body. This immersive visualization aids in making informed decisions during the drug design process.

iii. Virtual Clinical Trials: ABC Pharma takes advantage of the Metaverse to conduct virtual clinical trials. By creating virtual patient avatars, researchers can simulate the effects of their drug candidates on various patient populations. This allows them to evaluate the drug's efficacy and safety profiles without risking participant health. Virtual clinical trials save time and resources while providing valuable insights into the potential outcomes.

iv. Data Analytics and Machine Learning: ABC Pharma leverages the power of data analytics and machine learning algorithms within the Metaverse. By analyzing vast datasets from previous experiments, clinical trials, and scientific literature, researchers can identify patterns and correlations, leading to the discovery of novel drug targets or optimization strategies. Machine learning algorithms integrated into the Metaverse help in predicting drug properties and optimizing synthetic protocols.

Conclusion: Through the application of the Metaverse in drug discovery, ABC Pharma was able to streamline their drug development process significantly. The virtual laboratory simulations, collaborative research, molecular visualization, virtual clinical trials, and data analytics within the Metaverse enabled efficient decision-making and accelerated the identification of potential drug candidates. The Metaverse empowers pharmaceutical companies to expedite the discovery of life-saving drugs, ultimately benefiting patients worldwide.

d. Metaverse in Bio-chemistry

Biochemistry is a branch of science that explores the chemical processes and substances that occur within living organisms. It bridges the fields of biology and chemistry, focusing on the molecular and cellular mechanisms that underlie life.

Biochemistry plays a vital role in understanding the complexities of biological systems, from the structure and function of biomolecules like proteins, carbohydrates, lipids, and nucleic acids to the intricate metabolic pathways that regulate energy production and cellular processes. This knowledge is essential for advancing various fields, including medicine, agriculture, biotechnology, and environmental science. In medicine, biochemistry aids in the study of diseases, drug development, and personalized medicine. In agriculture, it contributes to improving crop yields and disease resistance. Biotechnological advancements, such as genetic engineering and gene editing, are also rooted in biochemistry. Overall, the importance of biochemistry lies in its fundamental contributions to unraveling the mysteries of life and its practical applications in enhancing human health and well-being.

The metaverse has immense applications in the field of biochemistry, offering new avenues for research, collaboration, and innovation. Here are a few examples of how the metaverse can be utilized in biochemistry:

o **Molecular Visualization and Simulation**: In the metaverse, biochemists can visualize and simulate complex molecular structures and interactions in a virtual environment. This allows for a deeper understanding of biological processes at the atomic and molecular level. By manipulating and observing virtual molecules, researchers can gain insights into biochemical reactions, protein folding, and drug-target interactions, aiding in the development of new therapies and treatments.

o **Collaborative Research and Data Sharing**: The metaverse provides a platform for biochemists to collaborate with colleagues from around the world. Virtual conferences, seminars, and research meetings can be conducted, facilitating the exchange of ideas, data, and findings. Through shared virtual laboratories, scientists can collaborate on experiments, share protocols, and collectively tackle complex biochemical challenges.

- **Drug Discovery and Design**: Virtual environments in the metaverse can be used for computer-aided drug design and virtual screening of potential drug candidates. Biochemists can simulate interactions between drugs and target molecules, predict binding affinities, and optimize drug structures virtually. This can significantly accelerate the drug discovery process, leading to the development of more effective and targeted therapies.

- **Protein Engineering and Design**: The metaverse can be utilized for virtual protein engineering and design. By leveraging computational tools and algorithms, biochemists can generate and optimize new protein structures with desired properties and functionalities. Virtual simulations can help assess protein stability, folding pathways, and protein-protein interactions, enabling the creation of novel enzymes, biocatalysts, and therapeutic proteins.

- **Education and Outreach**: The metaverse can serve as an immersive platform for biochemistry education and public outreach. Virtual classrooms, interactive tutorials, and educational games can be developed to engage students and the general public in learning about biochemical concepts and processes. This enhances accessibility and fosters enthusiasm for biochemistry among a broader audience. These applications demonstrate how the metaverse can transform the field of biochemistry, enabling advanced research, collaboration, and education. By harnessing virtual environments, biochemists can push the boundaries of knowledge, accelerate discoveries, and contribute to advancements in healthcare, biotechnology, and beyond.

o **VMD (Visual Molecular Dynamics)**: VMD is a molecular visualization and analysis software primarily used for the study of large biomolecular systems. It is popular among researchers working in the field of molecular dynamics simulations.

o **UCSF Chimera:** UCSF Chimera is a highly versatile molecular visualization and analysis program developed by the University of California, San Francisco. It is extensively used in the structural biology community.

o **Swiss-PdbViewer (DeepView):** Swiss-PdbViewer is a user-friendly and free molecular visualization software developed by the Swiss Institute of Bioinformatics (SIB).

o **Discovery Studio Visualizer:** Discovery Studio Visualizer is a free molecular visualization program provided by BIOVIA (formerly Accelrys) that offers powerful visualization and analysis tools for biomolecular structures.

o **Bioman biology**: Created and maintained by a biology teacher, Bioman Biology offers an outstanding selection of free biology interactives, games, quizzes, and virtual labs. All are standards-based and easy to use for students (no registration required). Teachers can track student progress by simply registering for their own account. Biology topics covered range from physiology to evolution to scientific methods.

o **Cell Homestasis Virtual Lab**: What happens to a living cell when it's placed in solutions of varying sugar concentrations? Use the virtual beakers, graduated cylinder, scale, dialysis tubes, and lab-grade sugar to find out. A simple experiment illustrating a key physiological principle.

- o **East Tennessee State University**: Great collection of higher-level free online biology labs covering ecology, evolution, and cell biology. Each experiment is available in both java and HTML5 - choose the HTML5 as java is no longer supported by most browsers. Users will enjoy changing parameters and observing the effects on honeybee foraging, male guppies' tail spots, or plant biodiversity. Perhaps most important in today's world: modeling and understanding the "tragedy of the commons" phenomenon.

- o **Online Dissection Resources:** Dissection Resources revolutionize learning by offering virtual exploration of anatomical structures. Through interactive 3D models, students can dissect and understand complex systems, fostering comprehensive understanding. These platforms provide ethical alternatives, flexibility, and cost-effectiveness, complementing traditional hands-on experiences for a well-rounded education.

- o **EON-XR:** EON-XR is a cutting-edge platform that offers immersive and interactive experiences in extended reality (XR) technology. XR encompasses virtual reality (VR), augmented reality (AR), and mixed reality (MR), providing users with a seamless blend of digital and real-world elements. EON-XR leverages this technology to deliver engaging and educational content across various industries, including education, training, and corporate sectors. With its user-friendly interface, EON-XR allows educators, trainers, and content creators to develop and deploy XR-based learning modules and simulations without extensive programming knowledge. This platform enables users to explore life like virtual environments, interact with 3D objects, and participate in hands-on training scenarios. Whether it's simulating complex procedures in medical training, providing interactive historical tours, or offering realistic safety drills, EON-XR opens up new possibilities for immersive learning experiences. By merging the physical and digital realms, EON-XR holds great potential to revolutionize how knowledge is acquired, retained, and applied, transforming traditional learning approaches into captivating and effective XR-enabled educational solutions.

- **FlyLab JS**: FlyLab JS is an interactive and web-based genetics simulator that allows users to experiment with genetic traits and inheritance patterns. Developed by the Virtual Genetics Lab at North Carolina State University, FlyLab JS is a powerful educational tool used to teach genetics concepts in a hands-on and engaging manner. With its user-friendly interface and customizable parameters, students and educators can explore various genetic scenarios and observe the outcomes, deepening their understanding of genetics principles.

- **HHMI BioInteractive**: HHMI BioInteractive is an exceptional educational resource developed by the Howard Hughes Medical Institute (HHMI). It offers a wealth of high-quality and engaging materials, including videos, animations, virtual labs, and interactive activities, to support biology and life science learning. Aimed at students, educators, and the general public, HHMI BioInteractive provides in-depth content on a wide range of topics, from genetics and evolution to ecology and neuroscience, making it a valuable tool for enhancing scientific understanding and promoting STEM education. Seven free virtual labs, designed for high school and college students, cover lizard and stickleback evolution, bacterial identification, clinical lab testing, neurophysiology, transgenic flies, and cardiology.

- **Learn.Genetics Virtual Labs**: From the University of Utah Genetic Science Learning Center, these five interactive labs cover techniques that are essential to any molecular biology lab. It provides a series of virtual laboratory simulations that allow users to explore various genetic concepts and experiments. With its user-friendly interface and engaging activities, Learn.Genetics Virtual Labs offers a hands-on learning experience for students, educators, and anyone interested in genetics, fostering a deeper understanding of genetic principles in a virtual and risk-free environment. With DNA and PCR becoming common vocabulary, it's a great introduction to highly topical technology.

e. Metaverse in Polymers chemistry

Polymers are large molecules composed of repeating subunits called monomers. They are an essential class of materials that have a wide range of applications in various industries. The importance of polymers lies in their versatility, lightweight nature, and cost-effectiveness. They are used in the production of plastics, fibers, rubber, adhesives, coatings, and many other products that we encounter daily. Polymers have revolutionized modern technology and have significantly contributed to advancements in medicine, electronics, packaging, and transportation. Moreover, their properties can be tailored to meet specific requirements, making them indispensable in diverse fields. Polymers play a vital role in the manufacturing of flexible displays, printed circuit boards (PCBs), and optoelectronic devices like organic light-emitting diodes (OLEDs). Flexible and lightweight polymers enable the production of bendable and foldable electronic gadgets, paving the way for the development of wearable technology and flexible electronics. Additionally, polymers serve as excellent insulators and dielectric materials, ensuring efficient power transmission and protection for sensitive electronic components. The integration of conductive polymers has facilitated the creation of printed electronics, where circuits and sensors can be directly printed onto flexible substrates.

Despite their significance, polymers also pose challenges, especially in terms of their impact on the environment. Many conventional polymers are non-biodegradable and can persist in the environment for centuries, leading to pollution and ecological harm. Addressing these challenges requires the development of more sustainable and eco-friendly polymers, as well as improving recycling and waste management practices. Additionally, the chemical composition and breakdown of polymers in certain applications can raise health concerns, emphasizing the need for comprehensive safety assessments. Through ongoing research and technological advancements, the responsible use and management of polymers can be achieved, striking a balance between their importance in modern society and environmental sustainability.

The metaverse holds immense potential for applications in the field of polymers, bringing new possibilities for research, design, and development of innovative polymer materials. Here are a few examples of how the metaverse can be utilized:

o **Virtual Polymer Modeling**: In the metaverse, researchers and polymer scientists can create virtual models of polymers with customizable properties and structures. Through simulations and molecular dynamics, they can study the behavior of polymers at the atomic and molecular level, gaining insights into their mechanical, thermal, and chemical properties. This virtual experimentation allows for rapid iteration and optimization of polymer designs, leading to the development of new materials with enhanced properties.

o **Collaborative Polymer Research**: The metaverse can serve as a collaborative platform for scientists and researchers from around the world to come together and share their knowledge and expertise in polymer science. Virtual conferences, seminars, and workshops can be hosted, fostering cross-disciplinary collaborations and accelerating advancements in the field. Researchers can exchange ideas, present their work, and collectively tackle complex challenges in polymer design and application.

o **Virtual Material Testing**: With the help of the metaverse, the testing and characterization of polymer materials can be simulated and visualized in a virtual environment. Researchers can virtually perform mechanical tests, measure properties such as tensile strength, elasticity, and deformation, and observe the response of polymers under different conditions. This digital testing can provide valuable insights for the development of optimized polymer materials for specific applications.

o **Sustainable Polymer Development**: The metaverse can contribute to the development of sustainable polymers by exploring virtual environments that simulate the life cycle of polymers, from production to disposal. Virtual assessments can be conducted to evaluate the environmental impact of different polymer formulations, allowing for the identification and optimization of eco-friendly alternatives. This promotes the development of more sustainable and recyclable polymer materials.

o **Virtual Manufacturing and Prototyping**: Virtual prototyping in polymer technology leverages advanced simulation tools to create and manipulate virtual models of polymer structures and materials. This approach enables researchers and engineers to predict material behaviors, test various formulations, and optimize designs without the need for physical prototypes. By streamlining development processes and reducing resource consumption, virtual prototyping accelerates innovation and enhances the efficiency of polymer technology advancements.

f. Metaverse in Food Science

Food science is a critical field that encompasses the study of the composition, processing, and safety of food products. It plays a vital role in ensuring the production of safe, nutritious, and high-quality food for consumers. Food scientists employ their knowledge in various areas, such as food chemistry, microbiology, sensory evaluation, and food engineering, to innovate and improve food products and processes. Understanding the molecular and chemical properties of food components allows for the development of functional foods that offer specific health benefits. Moreover, food science helps address global food challenges, such as food security, food waste reduction, and sustainable food production. Despite its importance, food science faces several challenges, including ensuring food safety in a rapidly changing global market, addressing issues related to food allergies and intolerances, and maintaining food quality during storage and transportation. Additionally, as consumer preferences evolve, food scientists must respond to demands for healthier and more environmentally friendly food options while navigating regulatory complexities. Meeting these challenges will require continuous research, collaboration, and innovation to ensure a safe and sustainable food supply for present and future generations.

The applications of the metaverse in food science are incredibly diverse and have the potential to revolutionize the way we research, develop, and experience food. Here are a few exciting examples:

- **Virtual Taste Testing**: In the metaverse, food scientists can conduct virtual taste tests, allowing them to analyze and evaluate the sensory aspects of food. By simulating taste, aroma, and texture, researchers can gather valuable insights into consumer preferences, optimize recipes, and develop new flavors. This virtual testing can reduce the need for physical product samples and provide rapid feedback for product development.

- **Nutritional Analysis and Customization**: The metaverse can offer virtual tools for analyzing the nutritional composition of different food formulations. By inputting ingredients and quantities into a virtual system, researchers can generate detailed nutritional profiles, including macronutrients, micronutrients, and allergen information. Furthermore, the metaverse can enable personalized nutrition recommendations, taking into account an individual's dietary needs, health goals, and preferences.

- **Sustainable Food Systems**: With the help of the metaverse, researchers and policymakers can model and simulate sustainable food systems. Virtual environments can allow for the analysis of the environmental impact of different farming, production, and distribution methods. This can inform decisions on optimizing resource utilization, reducing waste, and promoting sustainable practices in the food industry.

- **Culinary Education and Innovation**: Virtual culinary schools and innovation hubs in the metaverse can provide immersive learning experiences for aspiring chefs and food scientists. Virtual cooking classes, demonstrations, and workshops can be conducted, allowing participants to experiment with ingredients, techniques, and flavors in a virtual kitchen. This fosters creativity, collaboration, and the exploration of new culinary frontiers.

- **Food Safety and Quality Control**: The metaverse can facilitate virtual simulations of food safety protocols and quality control processes. By creating virtual food production environments, researchers can identify and address potential hazards, test food safety measures, and ensure compliance with regulations. This can enhance food safety practices and reduce risks for consumers. The metaverse has the potential to transform the field of food science by providing virtual platforms for research, experimentation, education, and innovation. It offers exciting opportunities to advance our understanding of food, create more sustainable and personalized food options, and enhance the overall food experience for consumers.

- **New Mexico State University Virtual Labs:** Very engaging eight virtual labs cover lab techniques of food science and are similar to real-world labs. They may not seem glamorous— but our food safety depends on these unheralded-yet-essential tests.

g. Metaverse in Environment Science

Environmental science is a multidisciplinary field that seeks to understand the interactions between the natural world and human activities, aiming to address environmental issues and promote sustainable practices. It encompasses the study of ecosystems, climate change, pollution, biodiversity, and more, with a focus on finding solutions to environmental challenges.

The application of the metaverse in environmental science presents exciting opportunities for research, education, and advocacy. In the metaverse, virtual environments can be created to simulate ecosystems, weather patterns, and environmental phenomena, allowing researchers to conduct experiments and analyze data in a controlled and immersive setting. The metaverse, with its immersive and interactive virtual environments, holds significant potential for various applications in environmental science. Some key applications of the metaverse in this field include:

i. **Ecosystem Modeling and Simulation**: The metaverse can be utilized to create virtual representations of ecosystems, allowing scientists to simulate and study complex interactions between species, climate patterns, and environmental factors. This enables researchers to gain insights into ecosystem dynamics, identify potential threats, and explore strategies for conservation and restoration.

ii. **Climate Change Research**: Metaverse-based climate models can help researchers visualize and understand the impacts of climate change on different regions and ecosystems. By incorporating real-time data and projections, scientists can conduct scenario-based simulations to assess the consequences of various climate change mitigation and adaptation strategies.

iii. **Environmental Monitoring and Data Visualization:** Virtual environments in the metaverse can serve as a platform for visualizing and analyzing environmental data from various sources. This allows researchers to monitor changes in environmental parameters, such as air and water quality, deforestation, and land use, in a more intuitive and accessible manner.

iv. Environmental Education and Outreach: The metaverse offers engaging and interactive tools for environmental education. Virtual field trips, interactive exhibits, and educational games can be designed to raise awareness about environmental issues and promote sustainability among students and the general public.

v. Virtual Reality Training for Environmental Professionals: Environmental professionals can use virtual reality (VR) simulations in the metaverse for training purposes. This includes practicing fieldwork, emergency response scenarios, and hazardous material handling in a safe and controlled virtual environment.

vi. Collaborative Research and Global Partnerships: The metaverse enables researchers from different geographic locations to collaborate seamlessly. Virtual meetings, conferences, and workshops facilitate the exchange of knowledge, data, and ideas, fostering global partnerships for addressing environmental challenges.

vii.Environmental Advocacy and Communication: Environmental organizations and activists can leverage the metaverse to create virtual campaigns and experiences that raise awareness about pressing environmental issues. These campaigns can engage a wider audience, promote eco-friendly behaviors, and encourage public support for conservation initiatives.

viii. Predictive Modeling and Conservation Planning: AI-powered metaverse applications can assist in predictive modeling for biodiversity conservation. By integrating data on species distributions, habitat suitability, and human impacts, researchers can develop conservation plans that prioritize areas for protection and restoration.

Overall, the applications of the metaverse in environmental science hold great promise for advancing research, education, and conservation efforts, contributing to a more sustainable and ecologically balanced future.

h. Metaverse in Forensic Science

Forensic science is a multidisciplinary field that utilizes scientific principles and methodologies to analyze and interpret evidence in criminal investigations. It plays a crucial role in the justice system, aiding law enforcement in solving crimes and supporting legal proceedings. Forensic analysis involves the examination of various types of evidence, such as DNA, fingerprints, ballistics, toxicology, digital data, and trace evidence. Through meticulous examination and analysis, forensic experts can establish links between suspects, victims, and crime scenes, helping to reconstruct the sequence of events and uncover critical details. The application of advanced technologies, such as DNA profiling and forensic databases, has significantly enhanced the accuracy and efficiency of forensic analysis. The results of forensic analysis are presented as expert testimony in court, contributing to the determination of guilt or innocence and ensuring justice is served. The combination of scientific rigor and analytical expertise in forensic science makes it an indispensable tool in solving complex criminal cases and ensuring the integrity of the criminal justice system.

The metaverse holds great potential for enhancing various aspects of forensic sciences. Here are a few exciting applications where the metaverse can contribute to this field:

i. **Virtual Crime Scene Reconstruction**: The metaverse can provide a platform for creating virtual crime scenes, allowing forensic investigators to reconstruct crime scenes digitally. This enables them to analyze and visualize evidence in a controlled and interactive environment, improving accuracy and efficiency in investigations.

ii. **Virtual Forensic Laboratories**: Within the metaverse, forensic scientists can have access to virtual laboratories equipped with the necessary tools and equipment for analyzing evidence. They can conduct virtual experiments, explore different forensic techniques, and practice analysis methods in a safe and cost-effective manner.

iii. **Training and Skill Development**: The metaverse can offer immersive training environments for forensic professionals. They can participate in virtual simulations and scenarios, practicing evidence collection, crime scene processing, and data analysis. This helps improve their skills, decision-making, and familiarity with various forensic procedures.

iv. **Collaborative Investigations**: Forensic scientists from different locations can collaborate virtually within the metaverse to work on complex cases. They can share evidence, compare findings, and collectively analyze data. This enhances collaboration, knowledge sharing, and cross-disciplinary expertise in forensic investigations.

v. **Virtual Autopsies and Anatomy Visualization**: The metaverse can provide realistic virtual autopsies and anatomical visualizations, allowing forensic pathologists and students to study human anatomy and examine post-mortem findings. This enhances understanding, training, and research opportunities in forensic pathology.

vi. Forensic Education and Outreach: The metaverse can facilitate interactive and engaging educational experiences in forensic sciences. Students and the general public can participate in virtual forensic workshops, solve virtual crime mysteries, and learn about forensic techniques and methodologies. This promotes awareness, interest, and understanding of forensic sciences. These applications of the metaverse in forensic sciences have the potential to enhance investigations, training, collaboration, and public engagement. It's an exciting prospect that can contribute to advancements in this critical field.

i. Metaverse in Fuel Science

Fuels, such as gasoline, diesel, natural gas, and biofuels, are critical energy sources that power various sectors of modern society. Their applications are diverse, with transportation being a significant consumer of fuels for cars, trucks, airplanes, and ships. Additionally, fuels are used in industrial processes, power generation, and heating applications. The convenience and energy density of fossil fuels have driven global economic development, but their combustion releases greenhouse gases, contributing to climate change. As a result, transitioning to cleaner and more sustainable fuels is a major challenge faced by the energy industry. The development and adoption of renewable energy sources and alternative fuels, such as electric vehicles, hydrogen, and advanced biofuels, are essential to reduce carbon emissions and combat environmental issues. Moreover, ensuring energy security, optimizing fuel efficiency, and managing fuel supply chains are ongoing challenges in the fuel industry. Striking a balance between meeting the world's energy demands and addressing environmental concerns remains a central focus for sustainable fuel applications.

The application of the metaverse in fuel technology opens up fascinating possibilities for exploring and advancing the energy sector. In this virtual realm, researchers, engineers, and energy experts can collaborate and experiment with innovative fuel sources and technologies, accelerating the development of sustainable and efficient energy solutions. Imagine a metaverse where scientists can simulate and test various fuel formulations, combustion processes, and energy storage systems in a virtual environment. They can observe the behavior of different fuels and their impact on performance, emissions, and overall energy efficiency. This virtual experimentation can enable rapid iteration and optimization of fuel compositions, leading to the discovery of cleaner and more sustainable energy sources. Furthermore, the metaverse can serve as a platform for virtual energy infrastructure planning and optimization. Engineers can design and simulate virtual power grids, evaluating the integration of renewable energy sources, battery storage systems, and smart grid technologies. This can help streamline the development of efficient and reliable energy networks in the real world. Additionally, the metaverse can facilitate education and public awareness about fuel technologies and their environmental impact. Virtual experiences and interactive simulations can help individuals understand the importance of clean energy and make informed decisions about their energy consumption habits. By harnessing the power of the metaverse, the fuel industry can benefit from accelerated research, improved collaboration, and enhanced public engagement, ultimately driving the transition to a more sustainable and environmentally friendly energy future.

j. Metaverse in Paint Technology

Paint technology plays a crucial role in various industries, from construction and automotive to aerospace and consumer goods. Its applications encompass aesthetic enhancement, corrosion protection, and surface functionalization. In the construction sector, paints are used to beautify buildings and provide weather resistance. In the automotive industry, they serve as protective coatings against rust and environmental elements. In aerospace, specialized paints ensure aerodynamic efficiency and protection against high-altitude conditions. However, paint technology faces challenges related to environmental impact and sustainability. Traditional paints may contain harmful volatile organic compounds (VOCs) that contribute to air pollution and health hazards. As a result, there is an increasing emphasis on developing eco-friendly, low-VOC, and water-based paints. Additionally, improving paint durability, color retention, and resistance to harsh conditions is an ongoing challenge. Advancements in nanotechnology and smart coatings offer potential solutions to enhance paint performance and address environmental concerns, making paint technology an area of continuous innovation and research.

The application of the metaverse in paint technology holds great potential for creativity and innovation. In this virtual realm, users can explore a wide range of possibilities for paint colors, textures, and finishes, unleashing their artistic talents without limitations. Imagine a metaverse where professional painters, interior designers, and even homeowners can virtually experiment with different paint colors and visualize them in real-time on digital representations of walls, furniture, or even entire rooms. This would eliminate the need for physical paint samples and provide a more accurate representation of how a particular color will look in a given space. In addition to color exploration, the metaverse could also offer advanced features like interactive tutorials for painting techniques, allowing both professionals and aspiring artists to learn and refine their skills in a virtual environment. Moreover, the metaverse could serve as a platform for collaboration among paint manufacturers, designers, and consumers. Manufacturers can showcase their entire range of paint products, allowing users to virtually browse and customize colors and finishes. Designers can collaborate with clients remotely, visually presenting different paint options and gathering feedback in real-time. Overall, the metaverse can revolutionize the paint industry by providing a dynamic and immersive platform for creativity, experimentation, and collaboration, ultimately enhancing the way we choose and apply paint in our physical world.

Paint and coatings brand Valspar announced the opening of the Valspar Color-verse™ web experience, a virtual house in the Metaverse featuring colors from the brand's recent 2023 Colors of the Year announcement. Visitors can interact with the 12 Colors of the Year, create their own art with the colors, and play a hyper-casual game, called the "Dash to DIY." experience enables visitors to demo and view the 12 featured colors in the Color-verse and get inspiration for their next project after experiencing the colors firsthand in the Metaverse. Visitors can navigate via pre-designated checkpoints or free roam using movement controls throughout the house.

k. Metaverse in Analytical Applications

The importance of analysis cannot be overstated in chemical industries, medicine industries, and research endeavors. In the chemical industry, analysis plays a pivotal role in ensuring product quality, process efficiency, and compliance with safety standards. Precise and accurate analysis helps identify chemical compositions, impurities, and potential hazards, enabling manufacturers to produce high-quality and safe products. In medicine industries, analysis is vital for drug development, clinical trials, and pharmaceutical quality control. Analyzing the chemical and biological properties of drugs helps determine their efficacy, safety, and interactions with the human body. In research, analysis is the foundation of scientific inquiry, aiding in the interpretation of data, formulation of hypotheses, and validation of research findings. Whether it's exploring new materials, understanding biological processes, or investigating environmental impacts, analysis drives innovation and advancements across these industries, leading to improved products, treatments, and knowledge.

The metaverse can bring several exciting applications to the field of analytical chemistry. Here are a few examples:

o **Virtual Analytical Instruments**: The metaverse can provide virtual representations of analytical instruments, such as mass spectrometers, chromatographs, and spectrophotometers. Researchers and students can interact with these instruments in a virtual environment, learning about their principles, operation, and data interpretation.

o **Simulated Data Analysis**: Within the metaverse, users can access simulated analytical data sets that mimic real-world scenarios. This allows researchers and students to practice data analysis techniques, such as peak identification, quantification, and statistical analysis, in a controlled and realistic virtual setting.

o **Collaborative Method Development**: The metaverse can facilitate collaborative method development in analytical chemistry. Researchers from different locations can come together virtually to share ideas, exchange data, and collectively develop and optimize analytical methods. This promotes knowledge sharing and advancements in the field.

o **Virtual Chemical Libraries**: Analytical chemists often work with extensive chemical libraries for compound identification and characterization. The metaverse can offer virtual chemical libraries that researchers can explore and search, providing information on molecular properties, spectra, and other relevant data for chemical analysis.

- o **Virtual Reality (VR) Visualization of Analytical Results:** By combining the metaverse with virtual reality (VR), researchers can visualize and interact with analytical results in immersive 3D environments. This allows for a deeper understanding of complex data sets, facilitating pattern recognition and insights that may be challenging to achieve through traditional 2D visualization methods.

- o **Remote Analytical Collaboration**: The metaverse enables remote collaboration among analytical chemists. Researchers can share data, compare results, and collaborate on projects without the need for physical presence. This promotes efficient collaboration across institutions and facilitates knowledge exchange. These applications of the metaverse in analytical chemistry empower researchers and students to explore, learn, and collaborate in innovative ways. They enhance understanding, foster creativity, and contribute to advancements in analytical techniques and methodologies.

l. Metaverse in Agriculture

Agricultural practices face a multitude of challenges that impact food production, sustainability, and the environment. Soil degradation, caused by erosion, overuse of chemical fertilizers, and loss of organic matter, threatens long-term soil health and fertility. Water scarcity and inefficient irrigation methods pose risks to crop yields, especially in regions facing climate change-induced droughts. Pests, diseases, and invasive species can devastate crops, requiring the use of pesticides, which may have environmental implications. Additionally, agricultural practices contribute to greenhouse gas emissions and can lead to deforestation, habitat destruction, and biodiversity loss. Finding sustainable solutions to these problems is crucial to ensure food security, preserve natural resources, and mitigate the impact of agriculture on the planet.

The metaverse can offer numerous applications in the fields of agricultural science and chemistry, helping to advance research, improve farming practices, and enhance sustainability. Here are ssome specific examples:

- **Virtual Farming Simulations:** The metaverse can provide virtual farming simulations where researchers, farmers, and students can experiment with different agricultural practices. They can explore virtual farms, test various cultivation techniques, and analyze the impact of factors such as soil composition, irrigation, and fertilization. This helps optimize crop yields, minimize resource inputs, and promote sustainable agriculture.

Farming Simulator is an immensely popular and realistic farming simulation video game series developed by Giants Software. Launched in 2008, the game quickly gained a dedicated fan base and has since evolved into one of the most successful simulation franchises.

One of the cornerstones of Farming Simulator's success is its active and passionate community. The official website acts as a platform where players can connect, share their experiences, and participate in discussions about the game.

Additionally, modding plays a significant role in enhancing the gameplay experience. The community creates and shares an extensive range of mods, which add new vehicles, equipment, maps, and gameplay mechanics to the base game, fostering a thriving and continuously evolving virtual farming world.

o **Precision Agriculture:** By integrating real-time sensor data with the metaverse, farmers can monitor crops and soil conditions in virtual environments. This allows for precise decision-making regarding irrigation, fertilization, and pest management. The metaverse can also assist in analyzing data trends to optimize farming practices and maximize productivity.

o **Crop Modeling and Prediction:** The metaverse can support the development of virtual crop models that simulate plant growth and predict yield under different environmental conditions. Researchers can incorporate data on weather patterns, soil quality, and genetic traits to forecast crop performance. This knowledge aids in crop selection, resource allocation, and risk management in agricultural systems. Interactive Education and Training: The metaverse can serve as a platform for interactive and immersive agricultural education. Students can engage in virtual field trips, participate in simulations of chemical analysis in crop production, and explore 3D models of plant anatomy and physiology. This enhances agricultural knowledge and prepares the next generation of scientists and farmers.

o **Environmental Impact Assessment:** Using the metaverse, researchers can evaluate the environmental impact of agricultural practices in a virtual setting. They can study factors like pesticide use, water consumption, and fertilizer runoff to assess their effects on ecosystems. This insight guides the development of sustainable agricultural strategies.

- o **Virtual Marketplaces and Supply Chains:** The metaverse can facilitate virtual marketplaces for farmers, suppliers, and buyers to connect and conduct business. It can enable tracking and verification of agricultural inputs, including chemical usage, seed varieties, and production methods, ensuring transparency and promoting responsible sourcing. These applications of the metaverse in agricultural science and chemistry have the potential to revolutionize farming practices, improve resource management, and foster sustainable agriculture. The integration of technology and scientific knowledge in these fields opens up new avenues for innovation and progress.

Following are a few examples of related softwares.

- o **The farming simulator https://www.farming-simulator.com/,** serves as the central hub for all things related to the Farming Simulator universe, offering players and fans a wealth of information, updates, and community engagement opportunities.
Key Features and Gameplay of Farming Simulator : Farming Simulator allows players to experience the joys and challenges of running a modern farm. The gameplay revolves around managing various aspects of farm life, such as crop cultivation, livestock rearing, forestry, and selling produce for profit. It offers a realistic depiction of agricultural activities, including operating a wide array of farming vehicles and machinery from renowned manufacturers like John Deere, Case IH, New Holland, and more.

Players start with a small plot of land and limited resources, but as they progress and generate income, they can expand their operations, purchase better equipment, and access larger fields. The game provides a peaceful and immersive experience where players can enjoy the tranquility of rural life while facing the dynamic challenges of weather, market fluctuations, and time management.

- **OneSoil's** is an innovative agricultural technology company that has made significant strides in transforming traditional farming practices through the power of data and artificial intelligence. With a vision to revolutionize agriculture, OneSoil's platform offers farmers access to cutting-edge tools and analytics that enable precision farming and data-driven decision-making. Leveraging satellite imagery and AI algorithms, OneSoil provides farmers with valuable insights into crop health, growth, and yield predictions, empowering them to optimize their operations and increase productivity. The digital scouting capabilities aid in early pest detection and effective pest management, while weather monitoring ensures farmers stay informed about changing conditions to plan and adapt accordingly. OneSoil's commitment to fostering an active agricultural community, coupled with their dedication to sustainability, positions them as a transformative force in the agricultural sector, helping farmers worldwide embrace technology-driven solutions for a more efficient and sustainable future.

- **Blue River Technology, accessible through its website http://www.bluerivertechnology.com,** stands at the forefront of agricultural innovation, harnessing the power of robotics and artificial intelligence (AI) to revolutionize modern farming practices. With a core focus on precision agriculture, Blue River's cutting-edge solutions empower farmers to make data-driven decisions through advanced imaging technology and machine learning algorithms. Their signature See & Spray technology sets a new standard for sustainable farming by enabling targeted herbicide application, reducing chemical usage, and promoting environmental stewardship. By seamlessly integrating robotics with existing agricultural equipment, Blue River ensures that farmers can easily adopt their transformative technologies, paving the way for a more efficient, productive, and sustainable future in the agriculture industry.

- **Prospera Technologies (https://www.prospera.ag/):** Prospera offers AI-driven solutions for crop monitoring and management. Their platform uses computer vision and machine learning algorithms to analyze crop health and provide actionable insights to farmers.

- **Taranis (https://www.taranis.ag/):** Taranis uses AI and CV to detect early signs of crop diseases, pests, and nutrient deficiencies. Their platform combines high-resolution imagery, weather data, and machine learning to support precision agriculture.

- **AgShift (https://www.agshift.com/):** AgShift employs AI and computer vision to assess and grade the quality of agricultural produce. Their platform helps in automating quality control processes and ensuring consistent grading standards.

- **The Climate Corporation (https://climate.com/):** The Climate Corporation, a subsidiary of Bayer, uses AI to provide farmers with field-level weather forecasts, agronomic insights, and personalized recommendations for planting and crop management.

- o **Agworld (https://www.agworld.com/):** Agworld integrates AI and data analytics to support farm management, enabling farmers to plan, execute, and track their agricultural activities efficiently.

- o **Gamaya (https://www.gamaya.com/):** Gamaya uses AI and hyperspectral imaging technology to monitor crop health and optimize resource allocation for precision farming.

- o **Ceres Imaging (https://www.ceresimaging.net/):** Ceres Imaging employs aerial imagery and AI algorithms to assess crop health and water stress, providing farmers with actionable data to optimize irrigation practices.

- o **Mavrx (https://www.mavrx.co/):** Mavrx utilizes AI and satellite imagery to offer data-driven insights for precision agriculture, assisting farmers in making informed decisions about their fields' health and performance.

- o **Ecoocean – Future Ocean's Online Game About Sustainable Fisheries**: EcoOcean is a unique online fishing simulator that aims to bring awareness to the global overfishing crisis while engaging users in a challenging sustainable fishing experiment. Easy to play and slightly addictive.

Please remember to visit the websites to explore their latest offerings and see how metaverse is evolving to support farm practices.

m. Metaverse in Averting accidents in Industry

Accidents in the chemical industry pose significant risks to human safety, the environment, and surrounding communities. Due to the nature of chemical processes and materials involved, accidents can lead to devastating consequences, including explosions, fires, toxic releases, and chemical spills. The potential impact on health, property, and ecosystems underscores the critical importance of stringent safety measures and adherence to regulations in this industry. Proper training, risk assessment, and maintenance protocols are essential to minimize the occurrence of accidents and ensure the safe handling and transportation of hazardous materials. Continuous efforts to improve safety practices and emergency response capabilities remain crucial in safeguarding workers and preventing incidents with far-reaching implications.

The metaverse can play a significant role in averting accidents in the chemical industry by providing innovative solutions for training, simulation, and safety measures. Here are a few applications where the metaverse can contribute to accident prevention:

i. **Virtual Training Simulations**: The metaverse can offer realistic virtual training simulations for chemical industry workers. They can practice emergency response procedures, hazardous material handling, and safety protocols in a virtual environment. This allows them to gain hands-on experience and develop critical skills without exposing themselves to real-life risks.

ii. **Immersive Hazard Identification:** Using the metaverse, workers can explore virtual chemical plants and identify potential hazards in an immersive 3D environment. They can learn to recognize and assess risks associated with different processes or equipment, enabling them to take preventive measures and mitigate potential accidents.

iii. **Virtual Safety Audits:** The metaverse can be used for virtual safety audits, where experts can remotely assess the safety protocols and systems of chemical facilities. They can identify gaps or areas for improvement and provide recommendations to enhance safety measures. This supports proactive accident prevention strategies.

iv. **Remote Monitoring and Maintenance:** Through the metaverse, technicians can remotely monitor chemical processes and equipment in real-time. Virtual sensors and data visualization can provide insights into potential malfunctions, leaks, or other safety concerns. This allows for timely intervention and preventive maintenance to avert accidents.

v. **Collaborative Safety Training:** The metaverse enables collaborative safety training among different stakeholders in the chemical industry, including workers, safety professionals, and management. They can participate in virtual workshops, share best practices, and collectively develop safety protocols and guidelines.

vi. **Augmented Reality (AR) Safety Assistance:** The metaverse, combined with AR technology, can provide on-site safety assistance to workers. AR overlays can display safety information, warnings, or instructions in real-time, ensuring that workers are aware of potential hazards and take appropriate precautions. By leveraging the metaverse's capabilities, the chemical industry can enhance safety awareness, training, and preventive measures, significantly reducing the risk of accidents. It enables a proactive approach to safety management, ultimately safeguarding both workers and the surrounding environment.

AI-based software like SAM GUARD revolutionizes workplace safety by harnessing the capabilities of artificial intelligence and computer vision technologies. SAM GUARD offers a proactive approach to accident prevention, continuously monitoring the workplace in real-time to detect potential safety hazards and risky behaviors. With its predictive analytics, the software identifies patterns and trends from historical data, enabling organizations to take preventive actions before accidents occur. SAM GUARD's ability to promptly alert relevant personnel and provide valuable insights ensures quick responses and informed decision-making. By integrating seamlessly with existing safety systems, the software creates a comprehensive safety ecosystem, enhancing workplace safety, reducing accidents, and fostering a culture of proactive risk management. SAM GUARD stands as a testament to the immense potential of AI in safeguarding employees and optimizing s afety protocols across industries.

5. Future of Metaverse

While we can't predict the future with certainty, here are a few potential directions the metaverse could take:

i. **Enhanced Immersion**: We can expect even more immersive and realistic experiences within the metaverse. Advancements in VR and haptic technologies might allow us to feel, touch, and interact with virtual objects as if they were real. (Haptic technologies, also known as haptics, encompass a fascinating field that revolves around creating tactile sensations and touch experiences through technology. By employing forces, vibrations, and motions, haptic devices can immerse users in an augmented reality of touch, adding a new dimension to digital interactions. These innovations find diverse applications, from virtual reality simulations, where users can feel and manipulate virtual objects, to enhancing the precision and control of teleoperation in fields like robotics and industrial automation. The term "haptic" derives from the Greek word "haptikos," meaning "pertaining to the sense of touch," which aptly reflects the essence of these technologies. As haptic technologies continue to evolve and become more accessible, they hold the potential to revolutionize how we interact with computers, devices, and even each other, paving the way for a more immersive and connected future.

ii. **Seamless Interconnectivity**: Seamless interconnectivity in the metaverse is the backbone of an immersive digital universe where multiple virtual worlds and experiences seamlessly blend together. This interconnectedness transcends boundaries, enabling users to navigate effortlessly between diverse environments, platforms, and applications. Imagine walking from a virtual shopping mall to a virtual concert venue without any interruptions. In the metaverse, individuals can seamlessly transition from a social gathering in one virtual space to a shared virtual workspace, or even dive into an epic gaming adventure, all without encountering jarring barriers or disruptions. This interconnected metaverse fosters a sense of continuity and fluidity, where users can carry their identities, assets, and experiences across different virtual realms. By promoting collaboration and communication on a scale never seen before, seamless interconnectivity in the metaverse has the potential to revolutionize social interactions, learning, entertainment, and business, reshaping the way we perceive and interact with the digital realm and each other. As technology progresses and more users participate, the metaverse's seamless interconnectivity will continue to evolve, opening up endless possibilities for the future of human interaction in the virtual space.

iii. **Creative Expression**: The metaverse will likely become a hub for creativity and self-expression. With improved tools and platforms, users may have endless possibilities to create and share their own virtual experiences, from virtual art galleries to interactive storytelling. Creative expression in the metaverse is an extraordinary avenue for individuals to unleash their imagination and bring their wildest ideas to life. Within this boundless digital realm, creators have the freedom to design and craft virtual worlds, artworks, avatars, and experiences that push the boundaries of what is possible in the physical world. Whether it's designing stunning virtual landscapes, composing original music for immersive environments, developing interactive storytelling experiences, or even creating unique fashion styles for avatars, the metaverse offers a playground for creative minds to thrive.

iv. **Social Collaboration**: Collaboration within the metaverse could become more dynamic and interactive. Imagine working with colleagues in a virtual office, attending virtual conferences, or even brainstorming in a virtual brainstorming room – all with the ability to seamlessly share ideas. Social events and entertainment may take on a whole new dimension, with concerts, art exhibitions, and gatherings hosted in virtual venues, reaching global audiences instantaneously. As the metaverse continues to evolve, the future of social collaboration within it is poised to reshape the very fabric of human connection, fostering a world where ideas, experiences, and cultures converge to build a more interconnected and inclusive global community.

v. **Virtual Economies:** Virtual economies might become more prominent, with the metaverse hosting thriving marketplaces for virtual goods, services, and experiences. Blockchain technology could ensure secure ownership and enable the creation of unique, tradable virtual assets. Of course, these are just some possibilities, and the future of the metaverse will likely hold many surprises. It's an exciting space that continues to evolve as technology advances and creative minds explore new frontiers.

6. Concluding remarks

In concluding this exploration of the metaverse, we have ventured into the boundless realm of virtual experiences and interconnected realities. From the inception of the concept to the rapid evolution of virtual worlds, we have witnessed the metaverse grow from a mere vision to a tangible reality that intertwines with our daily lives.

The metaverse has emerged as a powerful force, transforming the way we work, learn, play, and interact with one another. It has transcended the boundaries of distance and time, offering us opportunities to collaborate and connect like never before. Within its virtual landscapes, we have found new dimensions of creativity, expression, and understanding, enriching our lives in countless ways.

As the metaverse continues to evolve, we are faced with both immense possibilities and profound responsibilities. We must navigate the ethical, social, and technical challenges that arise from this brave new world. How we choose to shape the metaverse will define the path of human progress, as we seek to harness its potential for the betterment of society.

The journey to the metaverse is far from over. It is a perpetual voyage of discovery, innovation, and wonder. As pioneers of this digital frontier, we hold the power to mold this realm into one that reflects our highest aspirations and values.

Let us embrace the metaverse with curiosity, empathy, and a spirit of collaboration. Together, we can shape a future where the boundaries between the virtual and the physical dissolve, where creativity knows no bounds, and where humanity finds unity amidst the vastness of the metaverse.

With this final reflection, we invite you to embark on your own odyssey into the metaverse. May you continue to explore, create, and connect with boundless enthusiasm, for the metaverse is a canvas of dreams, waiting for you to paint your own masterpiece.

Thank you for joining us on this exhilarating journey.

As you turn the final page of "Unleashing the Metaverse: A Revolutionary Frontier in Chemistry and Related Fields," I invite you to reflect on the journey we've taken together. This book has explored the dynamic convergence of virtual realms and scientific frontiers, uncovering the transformative potential that emerges when technology meets the intricate world of chemistry and related fields. With each chapter, we've delved deeper into the possibilities, innovations, and collaborations in various fields that the metaverse offers to those seeking to push the boundaries of scientific exploration. As we conclude, I hope you're inspired to imagine the future possibilities that lie ahead—where the metaverse continues to empower us to forge new paths, challenge conventions, and reshape the very essence of discovery. Thank you for joining me on this intellectual odyssey, and may the insights within these pages fuel your curiosity and drive for years to come.

Dr. Manoj Bali (Ph.D. Chemistry),
Email - drmanojbali@gmail.com
Professor and Dean, University School of Sciences
Rayat Bahra University, V.P.O. Sahauran,
Tehsil Kharar, Distt. Mohali, Kharar, Punjab 140104 India

www.ingramcontent.com/pod-product-compliance
Lightning Source LLC
Chambersburg PA
CBHW062235290526
45794CB00006B/2295

* 9 7 9 8 8 6 0 3 8 7 4 4 7 *